JAN. 2011

Test Results for Mobile Device Acquisition Tool: Mobilyze 1.1

NCJ 232744

John H. Laub
Director, National Institute of Justice

This report was prepared for the National Institute of Justice, U.S. Department of Justice, by the Office of Law Enforcement Standards of the National Institute of Standards and Technology under Interagency Agreement 2003–IJ–R–029.

The National Institute of Justice is a component of the Office of Justice Programs, which also includes the Bureau of Justice Assistance, the Bureau of Justice Statistics, the Office of Juvenile Justice and Delinquency Prevention, and the Office for Victims of Crime.

January 2011

Test Results for Mobile Device Acquisition Tool:
Mobilyze Version 1.1

**National Institute of
Standards and Technology**
U.S. Department of Commerce

Contents

Introduction 1
How to Read This Report 1
1 Results Summary 2
2 Test Case Selection 2
3 Results by Test Assertion 4
    3.1 Acquisition Disruption 9
    3.2 PIM Data Acquisition 9
    3.3 Acquisition of Date/Time stamps for Text Messages 10
    3.4 Acquisition of Non-ASCII Characters 10
4 Testing Environment 10
    4.1 Test Computers 10
    4.2 Mobile Devices 10
    4.3 Internal Memory Data Objects 10
5 Test Results 12
    5.1 Test Results Report Key 12
    5.2 Test Details 12
        5.2.1 SPT–01 (iPhone 3Gs) 12
        5.2.2 SPT–02 (iPhone 3Gs) 14
        5.2.3 SPT–03 (iPhone 3Gs) 15
        5.2.4 SPT–04 (iPhone 3Gs) 16
        5.2.5 SPT–05 (iPhone 3Gs) 17
        5.2.6 SPT–06 (iPhone 3Gs) 18
        5.2.7 SPT–07 (iPhone 3Gs) 20
        5.2.8 SPT–08 (iPhone 3Gs) 21
        5.2.9 SPT–09 (iPhone 3Gs) 22
        5.2.10 SPT–10 (iPhone 3Gs) 23
        5.2.11 SPT–13 (iPhone 3Gs) 24
        5.2.12 SPT–24 (iPhone 3Gs) 25
        5.2.13 SPT–25 (iPhone 3Gs) 26
        5.2.14 SPT–29 (iPhone 3Gs) 27
        5.2.15 SPT–33 (iPhone 3Gs) 28
        5.2.16 SPT–38 (iPhone 3Gs) 29

Introduction

The Computer Forensics Tool Testing (CFTT) program is a joint project of the National Institute of Justice (NIJ), the department of Homeland Security (DHS), and the National Institute of Standards and Technology Office of Law Enforcement Standards (OLES) and Information Technology Laboratory (ITL). CFTT is supported by other organizations, including the Federal Bureau of Investigation, the U.S. Department of Defense Cyber Crime Center, U.S. Internal Revenue Service Criminal Investigation Division Electronic Crimes Program, and the U.S. Department of Homeland Security's Bureau of Immigration and Customs Enforcement, U.S. Customs and Border Protection and U.S. Secret Service. The objective of the CFTT program is to provide measurable assurance to practitioners, researchers, and other applicable users that the tools used in computer forensics investigations provide accurate results. Accomplishing this requires the development of specifications and test methods for computer forensics tools and subsequent testing of specific tools against those specifications.

Test results provide the information necessary for developers to improve tools, users to make informed choices, and the legal community and others to understand the tools' capabilities. This approach to testing computer forensic tools is based on well-recognized methodologies for conformance and quality testing. The specifications and test methods posted on the CFTT Web site (http://www.cftt.nist.gov/) are available for review and comment by the computer forensics community.

This document reports the results from testing Mobilyze, version 1.1, against the *Smart Phone Tool Test Assertions and Test Plan*, available at the CFTT Web site (www.cftt.nist.gov/mobile_devices.htm).

Test results from other software packages and the CFTT tool methodology can be found on NIJ's computer forensics tool testing Web page, http://www.ojp.usdoj.gov/nij/topics/technology/electronic-crime/cftt.htm.

How to Read This Report

This report is divided into five sections. The first section is a summary of the results from the test runs. This section is sufficient for most readers to assess the suitability of the tool for the intended use. The remaining sections of the report describe how the tests were conducted and provide documentation of test case run details that support the report summary. Sections 2 and 3 provide a justification for the selection of test cases and assertions from the set of possible cases that are defined in the test plan for smart phone forensic tools. The test cases are selected, in general, based on features offered by the tool. Section 4 lists the hardware and software used to run the test cases. Section 5 contains a description of each test case, test assertions used in the test case, the expected result and the actual result.

Test Results for Mobile Device Data Acquisition Tool

Tool Tested: Mobilyze
Version: 1.1

Run Environment: Mac OS X 10.5.4

Supplier: BlackBag Technologies, Inc.
Address: 300 Piercy Rd, San Jose, CA 95138

Tel: 408.844.8890
WWW: http://www.blackbagtech.com

1 Results Summary

Except for the following test cases: SPT–03, SPT–06, SPT–08, SPT–33, the tested tool acquired all supported data objects completely and accurately from the selected test mobile device (i.e., iPhone 3Gs). The exceptions were the following:

- Notification of device acquisition disruption was not successful. Test Case: SPT–03
- Maximum length address book entries reported in the preview-pane view were truncated. Test Case: SPT–06
- The delivery time for text messages displayed in the "Messages" tab are not reported. Test Case: SPT-08
- Non–ASCII address book entries and text messages are not properly reported in their native format. Test Case: SPT–33

2 Test Case Selection

Test cases used to test mobile device acquisition tools are defined in *Smart Phone Tool Test Assertions and Test Plan Version 1.0*. To test a tool, test cases are selected from the *Test Plan* document based on the features offered by the tool. Not all test cases or test assertions are appropriate for all tools. There is a core set of base cases that are executed for every tool tested. Tool features guide the selection of additional test cases. If a given tool implements a given feature then the test cases linked to that feature are run. Table 1a lists the test cases available in Mobilyze. Table 2a lists the test cases not available in Mobilyze.

Table 1a: Selected Test Cases (iPhone 3Gs)

Supported Test Cases	Cases Selected for Execution
Base Cases	SPT–01, SPT–02, SPT–03, SPT–04, SPT–05, SPT–06, SPT–07, SPT–08, SPT–09, SPT–10, SPT–13
Acquire mobile device internal memory and review reported data via supported generated report formats.	SPT–24

Supported Test Cases	Cases Selected for Execution
Acquire mobile device internal memory and review reported data via the preview pane.	SPT–25
After a successful mobile device internal memory, alter the case file via third-party means and attempt to re-open the case.	SPT–29
Acquire mobile device internal memory and review data containing non-ASCII characters.	SPT–33
Acquire mobile device internal memory and review hash values for vendor supported data objects.	SPT–38

Table 2a: Omitted Test Cases (iPhone 3Gs)

Unsupported Test Cases	Cases omitted – not executed
Acquire mobile device internal memory and review application related data (i.e., word documents, spreadsheet, presentation documents).	SPT–11
Acquire mobile device internal memory and review Internet related data (i.e., bookmarks, visited sites.	SPT–12
Acquire SIM memory over supported interfaces (e.g., PC/SC reader).	SPT–14
Attempt acquisition of a non-supported SIM.	SPT–15
Begin SIM acquisition and interrupt connectivity by interface disengagement.	SPT–16
Acquire SIM memory and review reported subscriber and equipment related information (i.e., SPN, ICCID, IMSI, MSISDN).	SPT–17
Acquire SIM memory and review reported Abbreviated Dialing Numbers (ADN).	SPT–18
Acquire SIM memory and review reported Last Numbers Dialed (LND).	SPT–19
Acquire SIM memory and review reported text messages (SMS, EMS).	SPT–20
Acquire SIM memory and review recoverable deleted text messages (SMS, EMS).	SPT–21
Acquire SIM memory and review reported location related data (i.e., LOCI, GPRSLOCI).	SPT–22
Acquire SIM memory by selecting a combination of supported data elements.	SPT–23
Acquire SIM memory and review reported data via supported generated report formats.	SPT–26
Acquire SIM memory and review reported data via the preview-pane.	SPT–27
Attempt acquisition of a password-protected SIM.	SPT–28
After a successful SIM acquisition, alter the case file via third-party means and attempt to re-open the case.	SPT–30
Perform a physical acquisition and review data output for readability.	SPT–31
Perform a physical acquisition and review reports for recoverable deleted data.	SPT–32

Unsupported Test Cases	Cases omitted – not executed
Acquire SIM memory and review data containing non-ASCII characters.	SPT–34
Begin acquisition on a PIN protected SIM to determine if the tool provides an accurate count of the remaining number of PIN attempts and if the PIN attempts are decremented when entering an incorrect value.	SPT–35
Begin acquisition on a SIM whose PIN attempts have been exhausted to determine if the tool provides an accurate count of the remaining number of PUK attempts and if the PUK attempts are decremented when entering an incorrect value.	SPT–36
Perform a stand-alone mobile device internal memory acquisition and review the status flags for text messages present on the SIM.	SPT–37
Acquire SIM memory and review hash values for vendor supported data objects.	SPT–39
Acquire mobile device internal memory and review data containing GPS longitude and latitude coordinates.	SPT–40

3 Results by Test Assertion

Table 3a summarizes the test results by assertion. The column labeled **Assertion** gives the text of each assertion. The column labeled **Tests** gives the number of test cases that use the given assertion. The column labeled **Anomaly** gives the section number in this report where the anomaly is discussed.

Table 3a: Assertions Tested: (iPhone 3Gs)

Assertions Tested	Tests	Anomaly
SPT–CA–01 If a cellular forensic tool provides support for connectivity of the target device then the tool shall successfully recognize the target device via all vendor supported interfaces (e.g., cable, Bluetooth, IrDA).	1	
SPT–CA–02 If a cellular forensic tool attempts to connect to a non-supported device then the tool shall notify the user that the device is not supported.	1	
SPT–CA–03 If connectivity between the mobile device and cellular forensic tool is disrupted then the tool shall notify the user that connectivity has been disrupted.	1	3.1
SPT–CA–04 If a cellular forensic tool completes acquisition of the target device without error then the tool shall have the ability to present acquired data objects in a useable format via either a preview-pane or generated report.	2	
SPT–CA–05 If a cellular forensic tool completes acquisition of the target device without error then subscriber-related information shall be presented in a useable format.	1	
SPT–CA–06 If a cellular forensic tool completes acquisition of the target device without error then equipment related information shall be	1	

Assertions Tested	Tests	Anomaly
presented in a useable format.		
SPT–CA–07 If a cellular forensic tool completes acquisition of the target device without error then address book entries shall be presented in a useable format.	1	
SPT–CA–08 If a cellular forensic tool completes acquisition of the target device without error then maximum length address book entries shall be presented in a useable format.	1	3.2
SPT–CA–09 If a cellular forensic tool completes acquisition of the target device without error then address book entries containing special characters shall be presented in a useable format.	1	
SPT–CA–10 If a cellular forensic tool completes acquisition of the target device without error then address book entries containing blank names shall be presented in a useable format.	1	
SPT–CA–11 If a cellular forensic tool completes acquisition of the target device without error then email addresses associated with address book entries shall be presented in a useable format.	1	
SPT–CA–12 If a cellular forensic tool completes acquisition of the target device without error then graphics associated with address book entries shall be presented in a useable format.	1	
SPT–CA–13 If a cellular forensic tool completes acquisition of the target device without error then datebook, calendar, note entries shall be presented in a useable format.	1	
SPT–CA–14 If a cellular forensic tool completes acquisition of the target device without error then maximum length datebook, calendar, note entries shall be presented in a useable format.	1	
SPT–CA–15 If a cellular forensic tool completes acquisition of the target device without error then call logs (incoming/outgoing/missed) shall be presented in a useable format.	1	
SPT–CA–16 If a cellular forensic tool completes acquisition of the target device without error then the corresponding date/time stamps and the duration of the call for call logs shall be presented in a useable format.	1	
SPT–CA–17 If a cellular forensic tool completes acquisition of the target device without error then ASCII text messages (i.e., SMS, EMS) shall be presented in a useable format.	1	
SPT–CA–18 If a cellular forensic tool completes acquisition of the target device without error then the corresponding date/time stamps for text messages shall be presented in a useable format.	1	3.3
SPT–CA–19 If a cellular forensic tool completes acquisition of the target device without error then the corresponding status (i.e., read, unread) for text messages shall be presented in a useable format.	1	
SPT–CA–20 If a cellular forensic tool completes acquisition of the target device without error then the corresponding sender / recipient phone numbers for text messages shall be presented in a useable format.	1	
SPT–CA–21 If a cellular forensic tool completes acquisition of the	1	

Assertions Tested	Tests	Anomaly
target device without error then MMS messages and associated audio shall be presented in a useable format.		
SPT–CA–22 If a cellular forensic tool completes acquisition of the target device without error then MMS messages and associated graphic files shall be presented in a useable format.	1	
SPT–CA–23 If a cellular forensic tool completes acquisition of the target device without error then MMS messages and associated video shall be presented in a useable format.	1	
SPT–CA–24 If a cellular forensic tool completes acquisition of the target device without error then stand-alone audio files shall be presented in a useable format via either an internal application or suggested third–party application.	1	
SPT–CA–25 If a cellular forensic tool completes acquisition of the target device without error then stand-alone graphic files shall be presented in a useable format via either an internal application or suggested third–party application.	1	
SPT–CA–26 If a cellular forensic tool completes acquisition of the target device without error then stand-alone video files shall be presented in a useable format via either an internal application or suggested third–party application.	1	
SPT–CA–29 If a cellular forensic tool provides the user with an "Acquire All" device data objects acquisition option then the tool shall complete the acquisition of all data objects without error.	2	
SPT–CA–32 If a cellular forensic tool completes two consecutive logical acquisitions of the target device without error then the payload (data objects) on the mobile device shall remain consistent.	1	
SPT–AO–25 If a cellular forensic tool completes acquisition of the target device without error then the tool shall present the acquired data in a useable format via supported generated report formats.	1	
SPT–AO–26 If a cellular forensic tool completes acquisition of the target device without error then the tool shall present the acquired data in a useable format in a preview-pane view.	1	3.2
SPT–AO–27 If the case file or individual data objects are modified via third-party means then the tool shall provide protection mechanisms disallowing or reporting data modification.	1	
SPT–AO–40 If the cellular forensic tool supports display of non-ASCII characters then the application should present address book entries in their native format.	1	3.4
SPT–AO–41 If the cellular forensic tool supports proper display of non-ASCII characters then the application should present text messages in their native format.	1	3.4
SPT–AO–43 If the cellular forensic tool supports hashing for individual data objects then the tool shall present the user with a hash value for each supported data object.	1	

Table 4a lists the assertions that were not tested, usually due to the tool not supporting an optional feature.

Table 4a: Assertions Not Tested (iPhone 3Gs)

Assertions Not Tested
SPT–CA–27 If a cellular forensic tool completes acquisition of the target device without error then device specific application related data shall be acquired and presented in a useable format via either an internal application or suggested third-party application.
SPT–CA–28 If a cellular forensic tool completes acquisition of the target device without error then Internet related data (i.e., bookmarks, visited sites) cached to the device shall be acquired and presented in a useable format.
SPT–CA–30 If a cellular forensic tool provides the user with a "Select All" individual device data objects then the tool shall complete the acquisition of all individually selected data objects without error.
SPT–CA–31 If a cellular forensic tool provides the user with the ability to "Select Individual" device data objects for acquisition then the tool shall acquire each exclusive data object without error.
SPT–AO–01 If a cellular forensic tool provides support for connectivity of the target SIM then the tool shall successfully recognize the target SIM via all tool-supported interfaces (e.g., PC/SC reader, proprietary reader, smart phone itself).
SPT–AO–02 If a cellular forensic tool attempts to connect to a non-supported SIM then the tool shall notify the user that the SIM is not supported.
SPT–AO–03 If a cellular forensic tool loses connectivity with the SIM reader then the tool shall notify the user that connectivity has been disrupted.
SPT–AO–04 If a cellular forensic tool completes acquisition of the target SIM without error then the SPN shall be presented in a useable format.
SPT–AO–05 If a cellular forensic tool completes acquisition of the target SIM without error then the ICCID shall be presented in a useable format.
SPT–AO–06 If a cellular forensic tool completes acquisition of the target SIM without error then the IMSI shall be presented in a useable format.
SPT–AO–07 If a cellular forensic tool completes acquisition of the target SIM without error then the MSISDN shall be presented in a useable format.
SPT–AO–08 If a cellular forensic tool completes acquisition of the target SIM without error then ASCII Abbreviated Dialing Numbers (ADN) shall be presented in a useable format.
SPT–AO–09 If a cellular forensic tool completes acquisition of the target SIM without error then maximum length ADNs shall be presented in a useable format.
SPT–AO–10 If a cellular forensic tool completes acquisition of the SIM without error then ADNs containing special characters shall be presented in a useable format.
SPT–AO–11 If a cellular forensic tool completes acquisition of the SIM without error then ADNs containing blank names shall be presented in a useable format.
SPT–AO–12 If a cellular forensic tool completes acquisition of the target SIM without error then Last Numbers Dialed (LND) shall be presented in a useable format.
SPT–AO–13 If a cellular forensic tool completes acquisition of the target SIM without

Assertions Not Tested
error then the corresponding date/time stamps for LNDs shall be presented in a useable format.
SPT–AO–14 If a cellular forensic tool completes acquisition of the target SIM without error then ASCII SMS text messages shall be presented in a useable format.
SPT–AO–15 If a cellular forensic tool completes acquisition of the target SIM without error then ASCII EMS text messages shall be presented in a useable format.
SPT–AO–16 If a cellular forensic tool completes acquisition of the target SIM without error then the corresponding date/time stamps for all text messages shall be presented in a useable format.
SPT–AO–17 If a cellular forensic tool completes acquisition of the target SIM without error then the corresponding status (i.e., read, unread) for text messages shall be presented in a useable format.
SPT–AO–18 If a cellular forensic tool completes acquisition of the target SIM without error then the corresponding sender / recipient phone numbers for text messages shall be presented in a useable format.
SPT–AO–19 If the cellular forensic tool completes acquisition of the target SIM without error then deleted text messages that have not been overwritten shall be presented in a useable format.
SPT–AO–20 If a cellular forensic tool completes acquisition of the target SIM without error then location related data (i.e., LOCI) shall be presented in a useable format.
SPT–AO–21 If a cellular forensic tool completes acquisition of the target SIM without error then location related data (i.e., GRPSLOCI) shall be presented in a useable format.
SPT–AO–22 If a cellular forensic tool provides the user with an "Acquire All" SIM data objects acquisition option then the tool shall complete the acquisition of all data objects without error.
SPT–AO–23 If a cellular forensic tool provides the user with a "Select All" individual SIM data objects then the tool shall complete the acquisition of all individually selected data objects without error.
SPT–AO–24 If a cellular forensic tool provides the user with the ability to "Select Individual" SIM data objects for acquisition then the tool shall acquire each exclusive data object without error.
SPT–AO–28 If the SIM is password–protected then the cellular forensic tool shall provide the examiner with the opportunity to input the PIN before acquisition.
SPT–AO–29 If a cellular forensic tool provides the examiner with the remaining number of authentication attempts then the application should provide an accurate count of the remaining PIN attempts.
SPT–AO–30 If a cellular forensic tool provides the examiner with the remaining number of PUK attempts then the application should provide an accurate count of the remaining PUK attempts.
SPT–AO–31 If the cellular forensic tool supports a physical acquisition of the target device then the tool shall complete the acquisition without error.
SPT–AO–32 If the cellular forensic tool supports the interpretation of address book entries present on the target device then the tool shall report recoverable active and deleted data or address book data remnants in a useable format.
SPT–AO–33 If the cellular forensic tool supports the interpretation of calendar, tasks, or

Assertions Not Tested
notes present on the target device then the tool shall report recoverable active and deleted calendar, tasks, or note data remnants in a useable format.
SPT–AO–34 If the cellular forensic tool supports the interpretation of call logs present on the target device then the tool shall report recoverable active and deleted call or call log data remnants in a useable format.
SPT–AO–35 If the cellular forensic tool supports the interpretation of SMS messages present on the target device then the tool shall report recoverable active and deleted SMS messages or SMS message data remnants in a useable format.
SPT–AO–36 If the cellular forensic tool supports the interpretation of EMS messages present on the target device then the tool shall report recoverable active and deleted EMS messages or EMS message data remnants in a useable format.
SPT–AO–37 If the cellular forensic tool supports the interpretation of audio files present on the target device then the tool shall report recoverable active and deleted audio data or audio file data remnants in a useable format.
SPT–AO–38 If the cellular forensic tool supports the interpretation of graphic files present on the target device then the tool shall report recoverable active and deleted graphic file data or graphic file data remnants in a useable format.
SPT–AO–39 If the cellular forensic tool supports the interpretation of video files present on the target device then the tool shall report recoverable active and deleted video file data or video file data remnants in a useable format.
SPT–AO–42 If the cellular forensic tool supports stand-alone acquisition of internal memory with the SIM present, then the contents of the SIM shall not be modified during internal memory acquisition.
SPT–AO–44 If the cellular forensic tool supports acquisition of GPS data then the tool shall present the user with the longitude and latitude coordinates for all GPS-related data in a useable format.

The following sections provide detailed information for the anomalies specified in Table 3a.

3.1 Acquisition Disruption

Notification of device acquisition disruption was not successful for test case SPT–03. The acquisition was disrupted by removing the cable from the mobile device during acquisition. When disrupting connectivity to the mobile device during acquisition, the Mobilyze application locks up and forces the user to force quit the application.

3.2 PIM Data Acquisition

Maximum length address book entries containing 126 characters were truncated after the 87th character in the preview-pane view for test case SPT–06.

3.3 Acquisition of Date/Time stamps for Text Messages

Time stamps for text messages were not reported for text messages displayed in the "Messages" tab for test case SPT-08.

3.4 Acquisition of Non-ASCII Characters

Acquisition of address book entries and text messages containing non-ASCII characters (i.e., French accent marks and Chinese characters) were not properly reported in their native format for test case SPT–33. Address book entries containing the following non-ASCII characters were reported as follows:

```
阿·哈拉       - displayed as: ÈòøÊÅ∂ÂìàÊãâ
Aurélien     - displayed as: Aur√©lien
```

Text messages containing the following non-ASCII characters were reported as follows:

```
阿·哈拉阿·哈拉  - displayed as: ÈòøÊÅ∂ÂìàÊãâ ÈòøÊÅ∂ÂìàÊãâ ÈòøÊÅ∂ÂìàÊãâ
ÀéïõÛ         - displayed as: √Ä√©√ø≈ç√ª
```

4 Testing Environment

The tests were run in the NIST CFTT lab. This section describes the testing environment including available computers, mobile devices and the data objects used to populate mobile devices and Subscriber Identity Modules.

4.1 Test Computers

One test computer was used.

p630542 has the following configuration:

MacBook Pro
Intel Core 2 Duo
Processor Speed: 2.6 GHz
Memory: 2 GB
Boot ROM Version: MBP31.0070.B05

4.2 Mobile Devices

The following table contains the mobile device used.

Make	Model	OS	Network
Apple iPhone	3Gs	iPhone	AT&T

4.3 Internal Memory Data Objects

The following data objects were used to populate the internal memory of the smart phones.

Data Objects	Data Elements
Address Book Entries	
	Regular Length
	Maximum Length
	Special Character
	Blank Name
	Regular Length, email
	Regular Length, graphic
	Deleted Entry
	Non-ASCII Entry
PIM Data	
	Regular Length
	Maximum Length
	Deleted Entry
	Special Character
Call Logs	
	Incoming
	Outgoing
	Missed
	Incoming – Deleted
	Outgoing – Deleted
	Missed – Deleted
Text Messages	
	Incoming SMS – Read
	Incoming SMS – Unread
	Outgoing SMS
	Incoming EMS – Read
	Incoming EMS – Unread
	Outgoing EMS
	Incoming SMS – Deleted
	Outgoing SMS – Deleted
	Incoming EMS – Deleted
	Outgoing EMS – Deleted
	Non–ASCII EMS
MMS Messages	
	Incoming Audio
	Incoming Graphic
	Incoming Video
	Outgoing Audio
	Outgoing Graphic
	Outgoing Video
Stand–alone data files	
	Audio
	Graphic
	Video

Data Objects	Data Elements
	Audio – Deleted
	Graphic – Deleted
	Video – Deleted
Application Data	
	Device Specific App Data
Location Data	
	GPS Coordinates

5 Test Results

The main item of interest for interpreting the test results is determining the conformance of the tool with the test assertions. Conformance with each assertion tested by a given test case is evaluated by examining the **Results** box of the test case details.

5.1 Test Results Report Key

A summary of the actual test results is presented in this report. The following table presents a description of each section of the test report summary.

Table 5 Test Results Report Key

Heading	Description
First Line:	Test case ID, name, and version of tool tested.
Case Summary:	Test case summary from *Smart Phone Tool Test Assertion and Test Plan*.
Assertions:	The test assertions applicable to the test case, selected from *Smart Phone Tool Test Assertion and Test Plan*.
Tester Name:	Name or initials of person executing test procedure.
Test Host:	Host computer executing the test.
Test Date:	Time and date that test was started.
Device:	Source mobile device, media (i.e., SIM).
Source Setup:	Acquisition interface.
Log Highlights:	Information extracted from various log files to illustrate conformance or non–conformance to the test assertions.
Results:	Expected and actual results for each assertion tested.
Analysis:	Whether or not the expected results were achieved.

5.2 Test Details

5.2.1 SPT–01 (iPhone 3Gs)

```
Test Case SPT-01 Mobilyze 1.1
Case         SPT-01 Acquire mobile device internal memory over tool-supported interfaces
Summary:     (e.g., cable, Bluetooth, IrDA).
Assertions:  SPT-CA-01 If a cellular forensic tool provides support for connectivity of
             the target device then the tool shall successfully recognize the target
```

Test Case SPT-01 Mobilyze 1.1	
	device via all vendor supported interfaces (e.g., cable, Bluetooth, IrDA). SPT-CA-04 If a cellular forensic tool completes acquisition of the target device without error then the tool shall have the ability to present acquired data objects in a useable format via either a preview-pane or generated report. SPT-CA-29 If a cellular forensic tool provides the user with an "Acquire All" device data objects acquisition option then the tool shall complete the acquisition of all data objects without error. SPT-CA-30 If a cellular forensic tool provides the user with a "Select All" individual device data objects then the tool shall complete the acquisition of all individually selected data objects without error. SPT-CA-31 If a cellular forensic tool provides the user with the ability to "Select Individual" device data objects for acquisition then the tool shall acquire each exclusive data object without error. SPT-CA-32 If a cellular forensic tool completes two consecutive logical acquisitions of the target device without error then the payload (data objects) on the mobile device shall remain consistent.
Tester Name:	rpa
Test Host:	P630542
Test Date:	Wed Aug 25 12:33:04 EDT 2010
Device:	iPhone3Gs
Source Setup:	OS: WIN XP Interface: cable
Log Highlights:	Created by Mobilyze 1.1 Acquisition started: Wed Aug 25 12:33:04 EDT 2010 Acquisition finished: Wed Aug 25 12:41:39 EDT 2010 Device connectivity was established via supported interface
Results:	

Assertion & Expected Result	Actual Result
SPT-CA-01 Device connectivity via supported interfaces.	as expected
SPT-CA-04 Readability and completeness of acquired data via supported reports.	as expected
SPT-CA-29 Acquire-All data objects acquisition.	as expected
SPT-CA-30 Select-All data objects acquisition.	as expected
SPT-CA-31 Select-Individual data objects acquisition.	as expected
SPT-CA-32 Perform back-to-back acquisitions, check device payload for modifications.	as expected

Analysis:	Expected results achieved

5.2.2 SPT-02 (iPhone 3Gs)

Test Case SPT-02 Mobilyze 1.1	
Case Summary:	SPT-02 Attempt internal memory acquisition of a non-supported mobile device.
Assertions:	SPT-CA-02 If a cellular forensic tool attempts to connect to a non-supported device then the tool shall notify the user that the device is not supported.
Tester Name:	rpa
Test Host:	P630542
Test Date:	Wed Aug 25 12:42:26 EDT 2010
Device:	nonsupported device
Source Setup:	OS: WIN XP Interface: cable
Log Highlights:	Created by Mobilyze 1.1 Acquisition started: Wed Aug 25 12:42:26 EDT 2010 Acquisition finished: Wed Aug 25 12:47:24 EDT 2010 Identification of non-supported devices was successful
Results:	<table><tr><th>Assertion & Expected Result</th><th>Actual Result</th></tr><tr><td>SPT-CA-02 Identification of non-supported devices.</td><td>as expected</td></tr></table>
Analysis:	Expected results achieved

5.2.3 SPT-03 (iPhone 3Gs)

Test Case SPT-03 Mobilyze 1.1	
Case Summary:	SPT-03 Begin mobile device internal memory acquisition and interrupt connectivity by interface disengagement.
Assertions:	SPT-CA-03 If connectivity between the mobile device and cellular forensic tool is disrupted then the tool shall notify the user that connectivity has been disrupted.
Tester Name:	rpa
Test Host:	P630542
Test Date:	Wed Aug 25 12:59:29 EDT 2010
Device:	iPhone3Gs
Source Setup:	OS: WIN XP Interface: cable
Log Highlights:	Created by Mobilyze 1.1 Acquisition started: Wed Aug 25 12:59:29 EDT 2010 Acquisition finished: Wed Aug 25 13:03:19 EDT 2010 Device acquisition disruption notification was not successful **Notes:** When disrupting connectivity to the mobile device being acquired, the Mobilyze application locks up and forces the user to issue a force quit.
Results:	<table><tr><th>Assertion & Expected Result</th><th>Actual Result</th></tr><tr><td>SPT-CA-03 Notification of device acquisition disruption.</td><td>Not as expected</td></tr></table>
Analysis:	Expected results Not achieved

5.2.4 SPT–04 (iPhone 3Gs)

Test Case SPT-04 Mobilyze 1.1	
Case Summary:	SPT-04 Acquire mobile device internal memory and review reported data via the preview-pane or generated reports for readability.
Assertions:	SPT-CA-04 If a cellular forensic tool completes acquisition of the target device without error then the tool shall have the ability to present acquired data objects in a useable format via either a preview-pane or generated report.
Tester Name:	rpa
Test Host:	P630542
Test Date:	Wed Aug 25 13:19:23 EDT 2010
Device:	iPhone3Gs
Source Setup:	OS: WIN XP Interface: cable
Log Highlights:	Created by Mobilyze 1.1 Acquisition started: Wed Aug 25 13:19:23 EDT 2010 Acquisition finished: Wed Aug 25 13:23:20 EDT 2010 Readability and completeness of acquired data was successful
Results:	<table><tr><th>Assertion & Expected Result</th><th>Actual Result</th></tr><tr><td>SPT-CA-04 Readability and completeness of acquired data via supported reports.</td><td>as expected</td></tr></table>
Analysis:	Expected results achieved

5.2.5 SPT–05 (iPhone 3Gs)

Test Case SPT-05 Mobilyze 1.1	
Case Summary:	SPT-05 Acquire mobile device internal memory and review reported subscriber and equipment related information (e.g., IMEI/MEID/ESN, MSISDN).
Assertions:	SPT-CA-05 If a cellular forensic tool completes acquisition of the target device without error then subscriber-related information shall be presented in a useable format. SPT-CA-06 If a cellular forensic tool completes acquisition of the target device without error then equipment related information shall be presented in a useable format.
Tester Name:	rpa
Test Host:	P630542
Test Date:	Wed Aug 25 13:27:43 EDT 2010
Device:	iPhone3Gs
Source Setup:	OS: WIN XP Interface: cable
Log Highlights:	Created by Mobilyze 1.1 Acquisition started: Wed Aug 25 13:27:43 EDT 2010 Acquisition finished: Wed Aug 25 13:39:26 EDT 2010 Subscriber and Equipment related data (i.e., MSISDN, IMEI) were acquired
Results:	<table><tr><th>Assertion & Expected Result</th><th>Actual Result</th></tr><tr><td>SPT-CA-05 Acquisition of MSISDN, IMSI.</td><td>as expected</td></tr><tr><td>SPT-CA-06 Acquisition of IMEI/MEID/ESN.</td><td>as expected</td></tr></table>
Analysis:	Expected results achieved

5.2.6 SPT–06 (iPhone 3Gs)

Test Case SPT-06 Mobilyze 1.1	
Case Summary:	SPT-06 Acquire mobile device internal memory and review reported PIM related data.
Assertions:	SPT-CA-07 If a cellular forensic tool completes acquisition of the target device without error then address book entries shall be presented in a useable format. SPT-CA-08 If a cellular forensic tool completes acquisition of the target device without error then maximum length address book entries shall be presented in a useable format. SPT-CA-09 If a cellular forensic tool completes acquisition of the target device without error then address book entries containing special characters shall be presented in a useable format. SPT-CA-10 If a cellular forensic tool completes acquisition of the target device without error then address book entries containing blank names shall be presented in a useable format. SPT-CA-11 If a cellular forensic tool completes acquisition of the target device without error then email addresses associated with address book entries shall be presented in a useable format. SPT-CA-12 If a cellular forensic tool completes acquisition of the target device without error then graphics associated with address book entries shall be presented in a useable format. SPT-CA-13 If a cellular forensic tool completes acquisition of the target device without error then datebook, calendar, note entries shall be presented in a useable format. SPT-CA-14 If a cellular forensic tool completes acquisition of the target device without error then maximum length datebook, calendar, note entries shall be presented in a useable format.
Tester Name:	rpa
Test Host:	P630542
Test Date:	Wed Aug 25 14:47:17 EDT 2010
Device:	iPhone3Gs
Source Setup:	OS: WIN XP Interface: cable
Log Highlights:	Created by Mobilyze 1.1 Acquisition started: Wed Aug 25 14:47:17 EDT 2010 Acquisition finished: Wed Aug 25 14:57:19 EDT 2010 Regular Length Address Book entries were acquired Maximum Length Address Book entries were partially acquired Special Character Address Book entries were acquired Blank Name Address Book entries were acquired Email addresses within Address Book entries were acquired Embedded graphics within Address Book entries were acquired ALL PIM related data was acquired - NA **Notes:** Maximum length address book entries containing 126 were truncated and only 87 characters were displayed in the preview-pane view. The generated report view contained all 126 characters of the maximum length address book entry.
Results:	

Assertion & Expected Result	Actual Result
SPT-CA-07 Acquisition of address book entries.	as expected
SPT-CA-08 Acquisition of maximum length address book entries.	partial
SPT-CA-09 Acquisition of address book entries containing special characters.	as expected
SPT-CA-10 Acquisition of address book entries containing a blank name entry.	as expected
SPT-CA-11 Acquisition of embedded email addresses within address book entries.	as expected
SPT-CA-12 Acquisition of embedded graphics within address book entries.	as expected

Test Case SPT-06 Mobilyze 1.1			
		SPT-CA-13 Acquisition of PIM data (i.e., datebook/calendar, notes).	NA
		SPT-CA-14 Acquisition of maximum length PIM data.	NA
	Analysis:	Partial results achieved	

5.2.7 SPT-07 (iPhone 3Gs)

Test Case SPT-07 Mobilyze 1.1	
Case Summary:	SPT-07 Acquire mobile device internal memory and review reported call logs.
Assertions:	SPT-CA-15 If a cellular forensic tool completes acquisition of the target device without error then call logs (incoming/outgoing/missed) shall be presented in a useable format. SPT-CA-16 If a cellular forensic tool completes acquisition of the target device without error then the corresponding date/time stamps and the duration of the call for call logs shall be presented in a useable format.
Tester Name:	rpa
Test Host:	P630542
Test Date:	Thu Aug 26 08:52:15 EDT 2010
Device:	iPhone3Gs
Source Setup:	OS: WIN XP Interface: cable
Log Highlights:	Created by Mobilyze 1.1 Acquisition started: Thu Aug 26 08:52:15 EDT 2010 Acquisition finished: Thu Aug 26 09:01:30 EDT 2010 All Call Logs (incoming, outgoing, missed) were acquired All Call Log date/time stamps data were correctly reported
Results:	<table><tr><th>Assertion & Expected Result</th><th>Actual Result</th></tr><tr><td>SPT-CA-15 Acquisition of call logs.</td><td>as expected</td></tr><tr><td>SPT-CA-16 Acquisition of call log date/time stamps.</td><td>as expected</td></tr></table>
Analysis:	Expected results achieved

5.2.8 SPT-08 (iPhone 3Gs)

Test Case SPT-08 Mobilyze 1.1	
Case Summary:	SPT-08 Acquire mobile device internal memory and review reported text messages.
Assertions:	SPT-CA-17 If a cellular forensic tool completes acquisition of the target device without error then ASCII text messages (i.e., SMS, EMS) shall be presented in a useable format. SPT-CA-18 If a cellular forensic tool completes acquisition of the target device without error then the corresponding date/time stamps for text messages shall be presented in a useable format. SPT-CA-19 If a cellular forensic tool completes acquisition of the target device without error then the corresponding status (i.e., read, unread) for text messages shall be presented in a useable format. SPT-CA-20 If a cellular forensic tool completes acquisition of the target device without error then the corresponding sender / recipient phone numbers for text messages shall be presented in a useable format.
Tester Name:	rpa
Test Host:	P630542
Test Date:	Thu Aug 26 09:02:43 EDT 2010
Device:	iPhone3Gs
Source Setup:	OS: WIN XP Interface: cable
Log Highlights:	Created by Mobilyze 1.1 Acquisition started: Thu Aug 26 09:02:43 EDT 2010 Acquisition finished: Thu Aug 26 09:15:20 EDT 2010 ALL text messages (SMS, EMS) were acquired Date/time stamps were partially reported for text messages Status flags were not reported for text messages - NA Sender and Recipient phone numbers associated with text messages were correctly reported **Notes**: The delivery time for text messages are not reported.
Results:	(see table below)
Analysis:	Partial results achieved

Assertion & Expected Result	Actual Result
SPT-CA-17 Acquisition of text messages.	as expected
SPT-CA-18 Acquisition of text message date/time stamps.	partial
SPT-CA-19 Acquisition of text message status flags.	as expected
SPT-CA-20 Acquisition of sender/recipient phone number associated with text messages.	as expected

5.2.9 SPT–09 (iPhone 3Gs)

Test Case SPT-09 Mobilyze 1.1	
Case Summary:	SPT-09 Acquire mobile device internal memory and review reported MMS multi-media related data (i.e., text, audio, graphics, video).
Assertions:	SPT-CA-21 If a cellular forensic tool completes acquisition of the target device without error then MMS messages and associated audio shall be presented in a useable format. SPT-CA-22 If a cellular forensic tool completes acquisition of the target device without error then MMS messages and associated graphic files shall be presented in a useable format. SPT-CA-23 If a cellular forensic tool completes acquisition of the target device without error then MMS messages and associated video shall be presented in a useable format.
Tester Name:	rpa
Test Host:	P630542
Test Date:	Thu Aug 26 09:24:20 EDT 2010
Device:	iPhone3Gs
Source Setup:	OS: WIN XP Interface: cable
Log Highlights:	Created by Mobilyze 1.1 Acquisition started: Thu Aug 26 09:24:20 EDT 2010 Acquisition finished: Thu Aug 26 09:26:19 EDT 2010 ALL MMS messages (Audio, Image, Video) were acquired **Notes**: The MMS attachments (i.e., audio, graphics, video files) are not linked to the associated textual portion of the message.
Results:	

Assertion & Expected Result	Actual Result
SPT-CA-21 Acquisition of audio MMS messages.	as expected
SPT-CA-22 Acquisition of graphic data image MMS messages.	as expected
SPT-CA-23 Acquisition of video MMS messages.	as expected

Analysis:	Expected results achieved

5.2.10 SPT–10 (iPhone 3Gs)

Test Case SPT-10 Mobilyze 1.1	
Case Summary:	SPT-10 Acquire mobile device internal memory and review reported stand-alone multi-media data (i.e., audio, graphics, video).
Assertions:	SPT-CA-24 If a cellular forensic tool completes acquisition of the target device without error then stand-alone audio files shall be presented in a useable format via either an internal application or suggested third-party application. SPT-CA-25 If a cellular forensic tool completes acquisition of the target device without error then stand-alone graphic files shall be presented in a useable format via either an internal application or suggested third-party application. SPT-CA-26 If a cellular forensic tool completes acquisition of the target device without error then stand-alone video files shall be presented in a useable format via either an internal application or suggested third-party application.
Tester Name:	rpa
Test Host:	P630542
Test Date:	Thu Aug 26 09:27:51 EDT 2010
Device:	iPhone3Gs
Source Setup:	OS: WIN XP Interface: cable
Log Highlights:	Created by Mobilyze 1.1 Acquisition started: Thu Aug 26 09:27:51 EDT 2010 Acquisition finished: Thu Aug 26 09:47:16 EDT 2010 ALL stand-alone data files (Audio, Image, Video) were acquired
Results:	<table><tr><th>Assertion & Expected Result</th><th>Actual Result</th></tr><tr><td>SPT-CA-24 Acquisition of stand-alone audio files.</td><td>as expected</td></tr><tr><td>SPT-CA-25 Acquisition of stand-alone graphic files.</td><td>as expected</td></tr><tr><td>SPT-CA-26 Acquisition of stand-alone video files.</td><td>as expected</td></tr></table>
Analysis:	Expected results achieved

5.2.11 SPT–13 (iPhone 3Gs)

Test Case SPT-13 Mobilyze 1.1	
Case Summary:	SPT-13 Acquire mobile device internal memory by selecting a combination of supported data elements.
Assertions:	SPT-CA-29 If a cellular forensic tool provides the user with an "Acquire All" device data objects acquisition option then the tool shall complete the acquisition of all data objects without error. SPT-CA-30 If a cellular forensic tool provides the user with an "Select All" individual device data objects then the tool shall complete the acquisition of all individually selected data objects without error. SPT-CA-31 If a cellular forensic tool provides the user with the ability to "Select Individual" device data objects for acquisition then the tool shall acquire each exclusive data object without error.
Tester Name:	rpa
Test Host:	P630542
Test Date:	Thu Aug 26 09:49:50 EDT 2010
Device:	iPhone3Gs
Source Setup:	OS: WIN XP Interface: cable
Log Highlights:	Created by Mobilyze 1.1 Acquisition started: Thu Aug 26 09:49:50 EDT 2010 Acquisition finished: Thu Aug 26 09:52:34 EDT 2010 Acquire All acquisition was successful
Results:	

Assertion & Expected Result	Actual Result
SPT-CA-29 Acquire-All data objects acquisition.	as expected
SPT-CA-30 Select-All data objects acquisition.	as expected
SPT-CA-31 Select-Individual data objects acquisition.	as expected

Analysis:	Expected results achieved

5.2.12 SPT-24 (iPhone 3Gs)

Test Case SPT-24 Mobilyze 1.1	
Case Summary:	SPT-24 Acquire mobile device internal memory and review reported data via supported generated report formats.
Assertions:	SPT-AO-25 If a cellular forensic tool completes acquisition of the target device without error then the tool shall present the acquired data in a useable format via supported generated report formats.
Tester Name:	rpa
Test Host:	P630542
Test Date:	Thu Aug 26 09:59:32 EDT 2010
Device:	iPhone3Gs
Source Setup:	OS: WIN XP Interface: cable
Log Highlights:	Created by Mobilyze 1.1 Acquisition started: Thu Aug 26 09:59:32 EDT 2010 Acquisition finished: Thu Aug 26 10:05:00 EDT 2010 Complete representation of known data via generated reports was successful
Results:	<table><tr><th>Assertion & Expected Result</th><th>Actual Result</th></tr><tr><td>SPT-AO-25 Comparison of known device data elements via generated reports.</td><td>as expected</td></tr></table>
Analysis:	Expected results achieved

5.2.13 SPT-25 (iPhone 3Gs)

Test Case SPT-25 Mobilyze 1.1	
Case Summary:	SPT-25 Acquire mobile device internal memory and review reported data via the preview pane.
Assertions:	SPT-AO-26 If a cellular forensic tool completes acquisition of the target device without error then the tool shall present the acquired data in a useable format in a preview-pane view.
Tester Name:	rpa
Test Host:	P630542
Test Date:	Thu Aug 26 10:05:23 EDT 2010
Device:	iPhone3Gs
Source Setup:	OS: WIN XP Interface: cable
Log Highlights:	Created by Mobilyze 1.1 Acquisition started: Thu Aug 26 10:05:23 EDT 2010 Acquisition finished: Thu Aug 26 10:13:36 EDT 2010 Complete representation of known data via preview-pane was not successful **Notes**: Maximum length address book entries were truncated in the preview-pane.
Results:	<table><tr><th>Assertion & Expected Result</th><th>Actual Result</th></tr><tr><td>SPT-AO-26 Comparison of known device data elements via preview-pane.</td><td>Not as expected</td></tr></table>
Analysis:	Expected results Not achieved

5.2.14 SPT-29 (iPhone 3Gs)

Test Case SPT-29 Mobilyze 1.1	
Case Summary:	SPT-29 After a successful mobile device internal memory acquisition, alter the case file via third-party means and attempt to re-open the case.
Assertions:	SPT-AO-27 If the case file or individual data objects are modified via third-party means then the tool shall provide protection mechanisms disallowing or reporting data modification.
Tester Name:	rpa
Test Host:	P630542
Test Date:	Thu Aug 26 10:14:07 EDT 2010
Device:	iPhone3Gs
Source Setup:	OS: WIN XP Interface: cable
Log Highlights:	Created by Mobilyze 1.1 Acquisition started: Thu Aug 26 10:14:07 EDT 2010 Acquisition finished: Thu Aug 26 10:15:36 EDT 2010 Notification of modified device memory data was successful
Results:	<table><tr><th>Assertion & Expected Result</th><th>Actual Result</th></tr><tr><td>SPT-AO-27 Notification of modified device case data.</td><td>as expected</td></tr></table>
Analysis:	Expected results achieved

5.2.15 SPT–33 (iPhone 3Gs)

Test Case SPT-33 Mobilyze 1.1	
Case Summary:	SPT-33 Acquire mobile device internal memory and review data containing non-ASCII characters.
Assertions:	SPT-AO-40 If the cellular forensic tool supports display of non-ASCII characters then the application should present address book entries in their native format. SPT-AO-41 If the cellular forensic tool supports proper display of non-ASCII characters then the application should present text messages in their native format.
Tester Name:	rpa
Test Host:	P630542
Test Date:	Thu Aug 26 10:16:48 EDT 2010
Device:	iPhone3Gs
Source Setup:	OS: WIN XP Interface: cable
Log Highlights:	Created by Mobilyze 1.1 Acquisition started: Thu Aug 26 10:16:48 EDT 2010 **Notes**: Address book entries containing Unicode characters were reported as follows: 阿•哈拉 - displayed as: ÈòøÊÅ∂ÂìàÊãâ Aurélien - displayed as: Aur√©lien Text messages containing Unicode characters were reported as follows: 阿•哈拉阿•哈拉 - displayed as: ÈòøÊÅ∂ÂìàÊãâ ÈòøÊÅ∂ÂìàÊãâ ÈòøÊÅ∂ÂìàÊãâ Àéïõû - displayed as: √Ä√©√Ø≈çv̂ª
Results:	

Assertion & Expected Result	Actual Result
SPT-AO-40 Acquisition of non-ASCII address book entries/ADNs.	as expected
SPT-AO-41 Acquisition of non-ASCII text messages.	as expected

Analysis:	Expected results achieved

5.2.16 SPT-38 (iPhone 3Gs)

Test Case SPT-38 Mobilyze 1.1	
Case Summary:	SPT-38 Acquire mobile device internal memory and review hash values for vendor supported data objects.
Assertions:	SPT-AO-43 If the cellular forensic tool supports hashing for individual data objects then the tool shall present the user with a hash value for each supported data object.
Tester Name:	rpa
Test Host:	P630542
Test Date:	Thu Aug 26 13:40:58 EDT 2010
Device:	iPhone3Gs
Source Setup:	OS: WIN XP Interface: cable
Log Highlights:	Created by Mobilyze 1.1 Acquisition started: Thu Aug 26 13:40:58 EDT 2010 Acquisition finished: Thu Aug 26 13:45:42 EDT 2010 Hash values were properly reported for individually acquired device data elements
Results:	<table><tr><th>Assertion & Expected Result</th><th>Actual Result</th></tr><tr><td>SPT-AO-43 Acquire data, check known hash values for consistency.</td><td>as expected</td></tr></table>
Analysis:	Expected results achieved

About the National Institute of Justice

A component of the Office of Justice Programs, NIJ is the research, development and evaluation agency of the U.S. Department of Justice. NIJ's mission is to advance scientific research, development and evaluation to enhance the administration of justice and public safety. NIJ's principal authorities are derived from the Omnibus Crime Control and Safe Streets Act of 1968, as amended (see 42 U.S.C. §§ 3721–3723).

The NIJ Director is appointed by the President and confirmed by the Senate. The Director establishes the Institute's objectives, guided by the priorities of the Office of Justice Programs, the U.S. Department of Justice, and the needs of the field. The Institute actively solicits the views of criminal justice and other professionals and researchers to inform its search for the knowledge and tools to guide policy and practice.

Strategic Goals

NIJ has seven strategic goals grouped into three categories:

Creating relevant knowledge and tools

1. Partner with state and local practitioners and policymakers to identify social science research and technology needs.
2. Create scientific, relevant, and reliable knowledge—with a particular emphasis on terrorism, violent crime, drugs and crime, cost-effectiveness, and community-based efforts—to enhance the administration of justice and public safety.
3. Develop affordable and effective tools and technologies to enhance the administration of justice and public safety.

Dissemination

4. Disseminate relevant knowledge and information to practitioners and policymakers in an understandable, timely and concise manner.
5. Act as an honest broker to identify the information, tools and technologies that respond to the needs of stakeholders.

Agency management

6. Practice fairness and openness in the research and development process.
7. Ensure professionalism, excellence, accountability, cost-effectiveness and integrity in the management and conduct of NIJ activities and programs.

Program Areas

In addressing these strategic challenges, the Institute is involved in the following program areas: crime control and prevention, including policing; drugs and crime; justice systems and offender behavior, including corrections; violence and victimization; communications and information technologies; critical incident response; investigative and forensic sciences, including DNA; less-than-lethal technologies; officer protection; education and training technologies; testing and standards; technology assistance to law enforcement and corrections agencies; field testing of promising programs; and international crime control.

In addition to sponsoring research and development and technology assistance, NIJ evaluates programs, policies, and technologies. NIJ communicates its research and evaluation findings through conferences and print and electronic media.

To find out more about the National Institute of Justice, please visit:

http://www.ojp.usdoj.gov/nij

or contact:

National Criminal Justice
  Reference Service
P.O. Box 6000
Rockville, MD 20849–6000
800–851–3420
http://www.ncjrs.gov

www.ingramcontent.com/pod-product-compliance
Lightning Source LLC
Chambersburg PA
CBHW081811170526
45167CB00008B/3394